THE

‖‖‖‖‖‖‖‖‖‖‖‖‖‖‖‖‖‖

I0459603

ART

OF

WAR

Ultimate Bilingual Edition (4-in-1)

English · Traditional Chinese · Simplified Chinese · Ancient Seal Script

SUN TZU

The Art of War: Ultimate Bilingual Edition (4-in-1)

Copyright © 2025 Jade Ink Press
All rights reserved.

This edition features:
- English translation by Lionel Giles (1910)
- Traditional Chinese text
- Simplified Chinese text
- Ancient Seal Script text

Original arrangement, presentation, and design by Jade Ink Press

Font: Chongxi Seal Script
Created by: Hsi-Jeng Hsieh, Hsin-Yi Wang, Hsü-Cheng Chi, Te-Ming Chuang
License: CC-BY-ND-3.0-TW-or-later
字體：崇義篆體
主創人員：謝清俊、王心怡、季旭昇、莊德明
授權：CC-BY-ND-3.0-TW-or-later

No part of this publication may be reproduced, stored in a retrieval system, or transmitted in any form or by any means, electronic, mechanical, photocopying, recording, or otherwise, without the prior written permission of the publisher.

Published by Jade Ink Press

First Edition: January 2025
Printed in the United States of America

Contents

ABOUT THIS EDITION

Welcome to this unique presentation of "The Art of War," a masterpiece that has guided leaders and strategists for over two millennia. In this edition, you'll experience Sun Tzu's wisdom in four different forms of writing, each offering its own window into history.

You'll find each passage presented in a thoughtfully designed two-page layout. On the left page, you'll see the Traditional Chinese characters at the top and Simplified Chinese at the bottom. On the right page, you'll find Lionel Giles' classic English translation at the top, with the ancient Small Seal Script at the bottom.

We've arranged the Traditional Chinese text from left to right and top to bottom, following modern reading conventions rather than the traditional right-to-left order. This choice makes the text more accessible to contemporary readers and allows for easier comparison with the English translation. We've also added modern punctuation to the Traditional Chinese text to enhance readability, though such marks were not used in ancient times. The Simplified Chinese version reflects the form commonly used in mainland China today.

For the English text, we use Lionel Giles' 1910 translation, widely respected for its clarity and accessibility.

The Small Seal Script appears at the bottom of the right page, maintaining its traditional right-to-left, top-to-bottom reading order without punctuation. Though this script was standardized by Emperor Ch'in Shih-huang-ti (the First Emperor of China) around 221 BCE—roughly three centuries after Sun Tzu's time—we chose to include it for its historical significance.

As China's first unified writing system, Small Seal Script represents a crucial link between ancient and modern Chinese writing. While Sun Tzu would have used an even earlier script, Small Seal Script preserves many characteristics of ancient Chinese writing, giving readers a sense of how the text might have appeared in antiquity.

Reading Small Seal Script follows traditional Chinese conventions: the text flows from right to left and top to bottom, without punctuation marks. While these ancient characters may appear artistic or decorative to modern eyes, each follows precise rules of composition that eventually evolved into today's Chinese characters.

Throughout this book, we've arranged the four versions side by side where possible. In some cases, where the English translation requires more space, the text continues onto the following page. We've focused solely on Sun Tzu's core text, presenting it cleanly without additional commentary or interpretation.

Whether you're a student of Chinese history, a language enthusiast, or simply curious about this timeless work, we hope this four-script edition helps you appreciate "The Art of War" in new and enlightening ways.

始計

孫子曰：兵者，國之大事，死生之地，存亡之道，不可不察也。

故經之以五事，校之以計，而索其情：一曰道，二曰天，三曰地，四曰將，五曰法。

始计

孙子曰：兵者，国之大事，死生之地，存亡之道，不可不察也。

故经之以五事，校之以计，而索其情：一曰道，二曰天，三曰地，四曰将，五曰法。

Laying Plans

Sun Tzu said: The art of war is of vital importance to the State.

It is a matter of life and death, a road either to safety or to ruin. Hence it is a subject of inquiry which can on no account be neglected.

The art of war, then, is governed by five constant factors, to be taken into account in one's deliberations, when seeking to determine the conditions obtaining in the field.

These are: (1) The Moral Law; (2) Heaven; (3) Earth; (4) The Commander; (5) Method and discipline.

SEAL SCRIPT

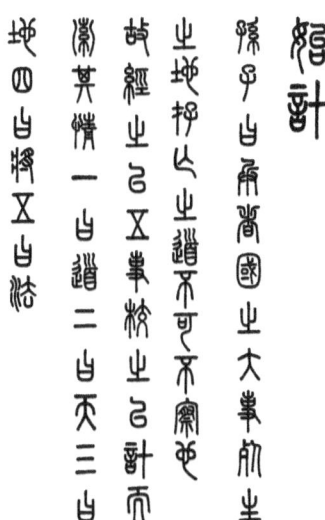

道者，令民與上同意，可與之死，可與之生，而不畏危也；天者，陰陽、寒暑、時制也；地者，遠近、險易、廣狹、死生也；將者，智、信、仁、勇、嚴也；法者，曲制、官道、主用也。凡此五者，將莫不聞，知之者勝，不知者不勝。

道者，令民与上同意，可与之死，可与之生，而不畏危也；天者，阴阳、寒暑、时制也；地者，远近、险易、广狭、死生也；将者，智、信、仁、勇、严也；法者，曲制、官道、主用也。凡此五者，将莫不闻，知之者胜，不知者不胜。

The Moral Law causes the people to be in complete accord with their ruler, so that they will follow him regardless of their lives, undismayed by any danger.

Heaven signifies night and day, cold and heat, times and seasons. Earth comprises distances, great and small; danger and security; open ground and narrow passes; the chances of life and death.

The Commander stands for the virtues of wisdom, sincerity, benevolence, courage and strictness.

By Method and discipline are to be understood the marshaling of the army in its proper subdivisions, the graduations of rank among the officers, the maintenance of roads by which supplies may reach the army, and the control of military expenditure.

These five heads should be familiar to every general: he who knows them will be victorious; he who knows them not will fail.

SEAL SCRIPT

故校之以計，而索其情，曰：主孰有道？將孰有能？天地孰得？法令孰行？兵眾孰強？士卒孰練？賞罰孰明？吾以此知勝負矣。將聽吾計，用之必勝，留之；將不聽吾計，用之必敗，去之。

故校之以计，而索其情，曰：主孰有道？将孰有能？天地孰得？法令孰行？兵众孰强？士卒孰练？赏罚孰明？吾以此知胜负矣。将听吾计，用之必胜，留之；将不听吾计，用之必败，去之。

Therefore, in your deliberations, when seeking to determine the military conditions, let them be made the basis of a comparison, in this wise:—

Which of the two sovereigns is imbued with the Moral law?

Which of the two generals has most ability?

With whom lie the advantages derived from Heaven and Earth?

On which side is discipline most rigorously enforced?

Which army is stronger?

On which side are officers and men more highly trained?

In which army is there the greater constancy both in reward and punishment?

By means of these seven considerations I can forecast victory or defeat.

The general that hearkens to my counsel and acts upon it, will conquer:—let such a one be retained in command! The general that hearkens not to my counsel nor acts upon it, will suffer defeat:—let such a one be dismissed!

SEAL SCRIPT

計利以聽，乃為之勢，以佐其外。勢者，因利而制權也。

兵者，詭道也。故能而示之不能，用而示之不用，近而示之遠，遠而示之近。利而誘之，亂而取之，實而備之，強而避之，怒而撓之，卑而驕之，佚而勞之，親而離之，攻其無備，出其不意。此兵家之勝，不可先傳也。

计利以听，乃为之势，以佐其外。势者，因利而制权也。

兵者，诡道也。故能而示之不能，用而示之不用，近而示之远，远而示之近。利而诱之，乱而取之，实而备之，强而避之，怒而挠之，卑而骄之，佚而劳之，亲而离之，攻其无备，出其不意。此兵家之胜，不可先传也。

While heeding the profit of my counsel, avail yourself also of any helpful circumstances over and beyond the ordinary rules.

According as circumstances are favourable, one should modify one's plans.

All warfare is based on deception.

Hence, when able to attack, we must seem unable; when using our forces, we must seem inactive; when we are near, we must make the enemy believe we are far away; when far away, we must make him believe we are near.

Hold out baits to entice the enemy. Feign disorder, and crush him.

If he is secure at all points, be prepared for him. If he is in superior strength, evade him.

If your opponent is of choleric temper, seek to irritate him. Pretend to be weak, that he may grow arrogant.

If he is taking his ease, give him no rest. If his forces are united, separate them.

Attack him where he is unprepared, appear where you are not expected.

These military devices, leading to victory, must not be divulged beforehand.

SEAL SCRIPT

13

夫未戰而廟算勝者，得算多也；未戰而廟算不勝者，得算少也。多算勝，少算不勝，而況無算乎！吾以此觀之，勝負見矣。

夫未战而庙算胜者，得算多也；未战而庙算不胜者，得算少也。多算胜，少算不胜，而况无算乎！吾以此观之，胜负见矣。

Now the general who wins a battle makes many calculations in his temple ere the battle is fought. The general who loses a battle makes but few calculations beforehand. Thus do many calculations lead to victory, and few calculations to defeat: how much more no calculation at all! It is by attention to this point that I can foresee who is likely to win or lose.

SEAL SCRIPT

作戰

孫子曰：凡用兵之法，馳車千駟，革車千乘，帶甲十萬，千里饋糧。則內外之費，賓客之用，膠漆之材，車甲之奉，日費千金，然後十萬之師舉矣。

作战

孙子曰：凡用兵之法，驰车千驷，革车千乘，带甲十万，千里馈粮。则内外之费，宾客之用，胶漆之材，车甲之奉，日费千金，然后十万之师举矣。

Waging War

Sun Tzu said: In the operations of war, where there are in the field a thousand swift chariots, as many heavy chariots, and a hundred thousand mail-clad soldiers, with provisions enough to carry them a thousand li, the expenditure at home and at the front, including entertainment of guests, small items such as glue and paint, and sums spent on chariots and armor, will reach the total of a thousand ounces of silver per day. Such is the cost of raising an army of 100,000 men.

SEAL SCRIPT

其用戰也，貴勝，久則鈍兵挫銳，攻城則力屈，久暴師則國用不足。夫鈍兵挫銳，屈力殫貨，則諸侯乘其弊而起，雖有智者，不能善其後矣。故兵聞拙速，未睹巧之久也。夫兵久而國利者，未之有也。故不盡知用兵之害者，則不能盡知用兵之利也。

其用战也，贵胜，久则钝兵挫锐，攻城则力屈，久暴师则国用不足。夫钝兵挫锐，屈力殚货，则诸侯乘其弊而起，虽有智者，不能善其后矣。故兵闻拙速，未睹巧之久也。夫兵久而国利者，未之有也。故不尽知用兵之害者，则不能尽知用兵之利也。

When you engage in actual fighting, if victory is long in coming, then men's weapons will grow dull and their ardor will be damped. If you lay siege to a town, you will exhaust your strength.

Again, if the campaign is protracted, the resources of the State will not be equal to the strain.

Now, when your weapons are dulled, your ardor damped, your strength exhausted and your treasure spent, other chieftains will spring up to take advantage of your extremity. Then no man, however wise, will be able to avert the consequences that must ensue.

Thus, though we have heard of stupid haste in war, cleverness has never been seen associated with long delays.

There is no instance of a country having benefited from prolonged warfare.

It is only one who is thoroughly acquainted with the evils of war that can thoroughly understand the profitable way of carrying it on.

SEAL SCRIPT

善用兵者，役不再籍，糧不三載，取用於國，因糧於敵，故軍食可足也。國之貧於師者遠輸，遠輸則百姓貧；近於師者貴賣，貴賣則百姓竭，財竭則急於丘役。力屈財殫，中原內虛於家，百姓之費，十去其七；公家之費，破軍罷馬，甲冑矢弩，戟楯矛櫓，丘牛大車，十去其六。

善用兵者，役不再籍，粮不三载，取用于国，因粮于敌，故军食可足也。国之贫于师者远输，远输则百姓贫；近于师者贵卖，贵卖则百姓竭，财竭则急于丘役。力屈财殚，中原内虚于家，百姓之费，十去其七；公家之费，破军罢马，甲冑矢弩，戟楯矛橹，丘牛大车，十去其六。

The skilful soldier does not raise a second levy, neither are his supply-wagons loaded more than twice.

Bring war material with you from home, but forage on the enemy. Thus the army will have food enough for its needs.

Poverty of the State exchequer causes an army to be maintained by contributions from a distance. Contributing to maintain an army at a distance causes the people to be impoverished.

On the other hand, the proximity of an army causes prices to go up; and high prices cause the people's substance to be drained away.

When their substance is drained away, the peasantry will be afflicted by heavy exactions.

With this loss of substance and exhaustion of strength, the homes of the people will be stripped bare, and three-tenths of their income will be dissipated; while government expenses for broken chariots, worn-out horses, breastplates and helmets, bows and arrows, spears and shields, protective mantles, draught-oxen and heavy wagons, will amount to four-tenths of its total revenue.

SEAL SCRIPT

故智將務食於敵，食敵一鍾，當吾二十鍾；
其秆一石，當吾二十石。故殺敵者，怒也；取敵
之利者，貨也。故車戰，得車十乘以上，賞其先
得者，而更其旌旗。車雜而乘之，卒善而養之，
是謂勝敵而益強。

故兵貴勝，不貴久。故知兵之將，民之司命。
國家安危之主也。

故智将务食于敌，食敌一钟，当吾二十钟；
其秆一石，当吾二十石。故杀敌者，怒也；取敌
之利者，货也。故车战，得车十乘以上，赏其先
得者，而更其旌旗。车杂而乘之，卒善而养之，
是谓胜敌而益强。

故兵贵胜，不贵久。故知兵之将，民之司命。
国家安危之主也。

Hence a wise general makes a point of foraging on the enemy. One cartload of the enemy's provisions is equivalent to twenty of one's own, and likewise a single picul of his provender is equivalent to twenty from one's own store.

Now in order to kill the enemy, our men must be roused to anger; that there may be advantage from defeating the enemy, they must have their rewards.

Therefore in chariot fighting, when ten or more chariots have been taken, those should be rewarded who took the first. Our own flags should be substituted for those of the enemy, and the chariots mingled and used in conjunction with ours. The captured soldiers should be kindly treated and kept.

This is called, using the conquered foe to augment one's own strength.

In war, then, let your great object be victory, not lengthy campaigns.

Thus it may be known that the leader of armies is the arbiter of the people's fate, the man on whom it depends whether the nation shall be in peace or in peril.

SEAL SCRIPT

23

謀攻

孫子曰：凡用兵之法，全國為上，破國次之；全軍為上，破軍次之；全旅為上，破旅次之；全卒為上，破卒次之；全伍為上，破伍次之。是故百戰百勝，非善之善者也；不戰而屈人之兵，善之善者也。

谋攻

孙子曰：凡用兵之法，全国为上，破国次之；全军为上，破军次之；全旅为上，破旅次之；全卒为上，破卒次之；全伍为上，破伍次之。是故百战百胜，非善之善者也；不战而屈人之兵，善之善者也。

Attack by Stratagem

Sun Tzu said: In the practical art of war, the best thing of all is to take the enemy's country whole and intact; to shatter and destroy it is not so good. So, too, it is better to capture an army entire than to destroy it, to capture a regiment, a detachment or a company entire than to destroy them.

Hence to fight and conquer in all your battles is not supreme excellence; supreme excellence consists in breaking the enemy's resistance without fighting.

SEAL SCRIPT

故上兵伐謀,其次伐交,其次伐兵,其下攻城。攻城之法,為不得已。修櫓轒轀,具器械,三月而後成;距闉,又三月而後已。將不勝其忿,而蟻附之,殺士三分之一,而城不拔者,此攻之災也。

故上兵伐谋,其次伐交,其次伐兵,其下攻城。攻城之法,为不得已。修橹轒轀,具器械,三月而后成;距闉,又三月而后已。将不胜其忿,而蚁附之,杀士三分之一,而城不拔者,此攻之灾也。

Thus the highest form of generalship is to baulk the enemy's plans; the next best is to prevent the junction of the enemy's forces; the next in order is to attack the enemy's army in the field; and the worst policy of all is to besiege walled cities.

The rule is, not to besiege walled cities if it can possibly be avoided. The preparation of mantlets, movable shelters, and various implements of war, will take up three whole months; and the piling up of mounds over against the walls will take three months more.

The general, unable to control his irritation, will launch his men to the assault like swarming ants, with the result that one-third of his men are slain, while the town still remains untaken. Such are the disastrous effects of a siege.

SEAL SCRIPT

故善用兵者，屈人之兵，而非戰也，拔人之城而非攻也，毀人之國而非久也，必以全爭於天下，故兵不頓而利可全，此謀攻之法也。

故用兵之法，十則圍之，五則攻之，倍則分之，敵則能戰之，少則能逃之，不若則能避之。故小敵之堅，大敵之擒也。

故善用兵者，屈人之兵，而非战也，拔人之城而非攻也，毁人之国而非久也，必以全争于天下，故兵不顿而利可全，此谋攻之法也。

故用兵之法，十则围之，五则攻之，倍则分之，敌则能战之，少则能逃之，不若则能避之。故小敌之坚，大敌之擒也。

Therefore the skilful leader subdues the enemy's troops without any fighting; he captures their cities without laying siege to them; he overthrows their kingdom without lengthy operations in the field.

With his forces intact he will dispute the mastery of the Empire, and thus, without losing a man, his triumph will be complete. This is the method of attacking by stratagem.

It is the rule in war, if our forces are ten to the enemy's one, to surround him; if five to one, to attack him; if twice as numerous, to divide our army into two.

If equally matched, we can offer battle; if slightly inferior in numbers, we can avoid the enemy; if quite unequal in every way, we can flee from him.

Hence, though an obstinate fight may be made by a small force, in the end it must be captured by the larger force.

SEAL SCRIPT

　　夫將者，國之輔也。輔周則國必強，輔隙則國必弱。故君之所以患於軍者三：不知軍之不可以進而謂之進，不知軍之不可以退而謂之退，是謂縻軍；不知三軍之事，而同三軍之政，則軍士惑矣；不知三軍之權，而同三軍之任，則軍士疑矣。三軍既惑且疑，則諸侯之難至矣。是謂亂軍引勝。

　　夫将者，国之辅也。辅周则国必强，辅隙则国必弱。故君之所以患于军者三：不知军之不可以进而谓之进，不知军之不可以退而谓之退，是谓縻军；不知三军之事，而同三军之政，则军士惑矣；不知三军之权，而同三军之任，则军士疑矣。三军既惑且疑，则诸侯之难至矣。是谓乱军引胜。

Now the general is the bulwark of the State; if the bulwark is complete at all points; the State will be strong; if the bulwark is defective, the State will be weak.

There are three ways in which a ruler can bring misfortune upon his army:—

By commanding the army to advance or to retreat, being ignorant of the fact that it cannot obey. This is called hobbling the army.

By attempting to govern an army in the same way as he administers a kingdom, being ignorant of the conditions which obtain in an army. This causes restlessness in the soldier's minds.

By employing the officers of his army without discrimination, through ignorance of the military principle of adaptation to circumstances. This shakes the confidence of the soldiers.

But when the army is restless and distrustful, trouble is sure to come from the other feudal princes. This is simply bringing anarchy into the army, and flinging victory away.

SEAL SCRIPT

31

故知勝有五：知可以戰與不可以戰者，勝。識眾寡之用者，勝。上下同欲者，勝。以虞待不虞者，勝。將能而君不御者，勝。此五者，知勝之道也。

故曰：知己知彼，百戰不貽；不知彼而知己，一勝一負；不知彼不知己，每戰必敗。

故知胜有五：知可以战与不可以战者，胜。识众寡之用者，胜。上下同欲者，胜。以虞待不虞者，胜。将能而君不御者，胜。此五者，知胜之道也。

故曰：知己知彼，百战不贻；不知彼而知己，一胜一负；不知彼不知己，每战必败。

Thus we may know that there are five essentials for victory:

He will win who knows when to fight and when not to fight.

He will win who knows how to handle both superior and inferior forces.

He will win whose army is animated by the same spirit throughout all its ranks.

He will win who, prepared himself, waits to take the enemy unprepared.

He will win who has military capacity and is not interfered with by the sovereign.

Victory lies in the knowledge of these five points.

Hence the saying: If you know the enemy and know yourself, you need not fear the result of a hundred battles. If you know yourself but not the enemy, for every victory gained you will also suffer a defeat. If you know neither the enemy nor yourself, you will succumb in every battle.

SEAL SCRIPT

33

軍形

孫子曰：昔之善戰者，先為不可勝，以待敵之可勝。不可勝在己，可勝在敵。故善戰者，能為不可勝，不能使敵必可勝。故曰：勝可知，而不可為。

不可勝者，守也；可勝者，攻也。守則不足，攻則有餘。善守者，藏於九地之下，善攻者，動於九天之上，故能自保而全勝也。

军形

孙子曰：昔之善战者，先为不可胜，以待敌之可胜。不可胜在己，可胜在敌。故善战者，能为不可胜，不能使敌必可胜。故曰：胜可知，而不可为。

不可胜者，守也；可胜者，攻也。守则不足，攻则有余。善守者，藏于九地之下，善攻者，动于九天之上，故能自保而全胜也。

Tactical Dispositions

Sun Tzu said: The good fighters of old first put themselves beyond the possibility of defeat, and then waited for an opportunity of defeating the enemy.

To secure ourselves against defeat lies in our own hands, but the opportunity of defeating the enemy is provided by the enemy himself.

Thus the good fighter is able to secure himself against defeat, but cannot make certain of defeating the enemy.

Hence the saying: One may know how to conquer without being able to do it.

Security against defeat implies defensive tactics; ability to defeat the enemy means taking the offensive.

Standing on the defensive indicates insufficient strength; attacking, a superabundance of strength.

The general who is skilled in defence hides in the most secret recesses of the earth; he who is skilled in attack flashes forth from the topmost heights of heaven. Thus on the one hand we have ability to protect ourselves; on the other, a victory that is complete.

SEAL SCRIPT

35

見勝不過眾人之所知，非善之善者也；戰勝而天下曰善，非善之善者也。故舉秋毫不為多力，見日月不為明目，聞雷霆不為聰耳。古之善戰者，勝於易勝者也。故善戰者之勝也，無智名，無勇功，故其戰勝不忒。不忒者，其所措必勝，勝已敗者也。故善戰者，先立於不敗之地，而不失敵之敗也。是故勝兵先勝，而後求戰，敗兵先戰而後求勝。

见胜不过众人之所知，非善之善者也；战胜而天下曰善，非善之善者也。故举秋毫不为多力，见日月不为明目，闻雷霆不为聪耳。古之善战者，胜于易胜者也。故善战者之胜也，无智名，无勇功，故其战胜不忒。不忒者，其所措必胜，胜已败者也。故善战者，先立于不败之地，而不失敌之败也。是故胜兵先胜，而后求战，败兵先战而后求胜。

To see victory only when it is within the ken of the common herd is not the acme of excellence.

Neither is it the acme of excellence if you fight and conquer and the whole Empire says, "Well done!"

To lift an autumn hair is no sign of great strength; to see the sun and moon is no sign of sharp sight; to hear the noise of thunder is no sign of a quick ear.

What the ancients called a clever fighter is one who not only wins, but excels in winning with ease.

Hence his victories bring him neither reputation for wisdom nor credit for courage.

He wins his battles by making no mistakes. Making no mistakes is what establishes the certainty of victory, for it means conquering an enemy that is already defeated.

Hence the skilful fighter puts himself into a position which makes defeat impossible, and does not miss the moment for defeating the enemy.

Thus it is that in war the victorious strategist only seeks battle after the victory has been won, whereas he who is destined to defeat first fights and afterwards looks for victory.

SEAL SCRIPT

善用兵者，修道而保法，故能為勝敗之政。兵法：一曰度，二曰量，三曰數，四曰稱，五曰勝。地生度，度生量，量生數，數生稱，稱生勝。故勝兵若以鎰稱銖，敗兵若以銖稱鎰。勝者之戰，若決積水於千仞之谿者，形也。

善用兵者，修道而保法，故能为胜败之政。兵法：一曰度，二曰量，三曰数，四曰称，五曰胜。地生度，度生量，量生数，数生称，称生胜。故胜兵若以镒称铢，败兵若以铢称镒。胜者之战，若决积水于千仞之谿者，形也。

The consummate leader cultivates the moral law, and strictly adheres to method and discipline; thus it is in his power to control success.

In respect of military method, we have, firstly, Measurement; secondly, Estimation of quantity; thirdly, Calculation; fourthly, Balancing of chances; fifthly, Victory.

Measurement owes its existence to Earth; Estimation of quantity to Measurement; Calculation to Estimation of quantity; Balancing of chances to Calculation; and Victory to Balancing of chances.

A victorious army opposed to a routed one, is as a pound's weight placed in the scale against a single grain.

The onrush of a conquering force is like the bursting of pent-up waters into a chasm a thousand fathoms deep.

SEAL SCRIPT

39

兵勢

孫子曰：凡治眾如治寡，分數是也；鬥眾如鬥寡，形名是也；三軍之眾，可使必受敵而無敗者，奇正是也；兵之所加，如以碬投卵者，虛實是也。

兵势

孙子曰：凡治众如治寡，分数是也；斗众如斗寡，形名是也；三军之众，可使必受敌而无败者，奇正是也；兵之所加，如以碬投卵者，虚实是也。

Energy

Sun Tzu said: The control of a large force is the same principle as the control of a few men: it is merely a question of dividing up their numbers.

Fighting with a large army under your command is nowise different from fighting with a small one: it is merely a question of instituting signs and signals.

To ensure that your whole host may withstand the brunt of the enemy's attack and remain unshaken—this is effected by manoeuvres direct and indirect.

That the impact of your army may be like a grindstone dashed against an egg—this is effected by the science of weak points and strong.

SEAL SCRIPT

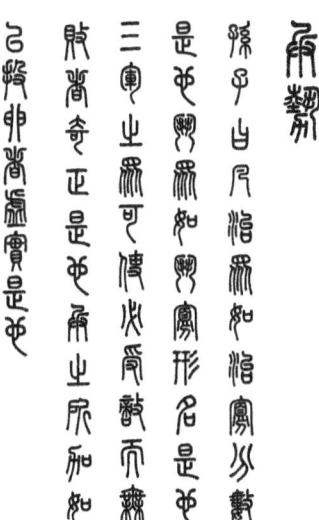

凡戰者，以正合，以奇勝。故善出奇者，無窮如天地，不竭如江海。終而複始，日月是也。死而復生，四時是也。聲不過五,五聲之變，不可勝聽也；色不過五,五色之變，不可勝觀也；味不過五,五味之變，不可勝嘗也；戰勢，不過奇正，奇正之變，不可勝窮也。奇正相生，如循環之無端，熟能窮之哉？

凡战者，以正合，以奇胜。故善出奇者，无穷如天地，不竭如江海。终而复始，日月是也。死而复生，四时是也。声不过五,五声之变，不可胜听也；色不过五,五色之变，不可胜观也；味不过五,五味之变，不可胜尝也；战势，不过奇正，奇正之变，不可胜穷也。奇正相生，如循环之无端，熟能穷之哉？

In all fighting, the direct method may be used for joining battle, but indirect methods will be needed in order to secure victory.

Indirect tactics, efficiently applied, are inexhaustible as Heaven and Earth, unending as the flow of rivers and streams; like the sun and moon, they end but to begin anew; like the four seasons, they pass away to return once more.

There are not more than five musical notes, yet the combinations of these five give rise to more melodies than can ever be heard.

There are not more than five primary colors, yet in combination they produce more hues than can ever been seen.

There are not more than five cardinal tastes, yet combinations of them yield more flavors than can ever be tasted.

In battle, there are not more than two methods of attack—the direct and the indirect; yet these two in combination give rise to an endless series of manoeuvres.

The direct and the indirect lead on to each other in turn. It is like moving in a circle—you never come to an end. Who can exhaust the possibilities of their combination?

SEAL SCRIPT

43

激水之疾，至於漂石者，勢也；鷙鳥之疾，至於毀折者，節也。是故善戰者，其勢險，其節短。勢如張弩，節如發機。

激水之疾，至于漂石者，势也；鸷鸟之疾，至于毁折者，节也。是故善战者，其势险，其节短。势如张弩，节如发机。

The onset of troops is like the rush of a torrent which will even roll stones along in its course.

The quality of decision is like the well-timed swoop of a falcon which enables it to strike and destroy its victim.

Therefore the good fighter will be terrible in his onset, and prompt in his decision.

Energy may be likened to the bending of a crossbow; decision, to the releasing of a trigger.

SEAL SCRIPT

45

紛紛紜紜，鬥亂而不可亂也；渾渾沌沌，形圓而不可敗也。亂生於治，怯生於勇，弱生於強。治亂，數也；勇怯，勢也；強弱，形也。故善動敵者，形之，敵必從之；予之，敵必取之。以利動之，以卒待之。

纷纷纭纭，斗乱而不可乱也；浑浑沌沌，形圆而不可败也。乱生于治，怯生于勇，弱生于强。治乱，数也；勇怯，势也；强弱，形也。故善动敌者，形之，敌必从之；予之，敌必取之。以利动之，以卒待之。

Amid the turmoil and tumult of battle, there may be seeming disorder and yet no real disorder at all; amid confusion and chaos, your array may be without head or tail, yet it will be proof against defeat.

Simulated disorder postulates perfect discipline, simulated fear postulates courage; simulated weakness postulates strength.

Hiding order beneath the cloak of disorder is simply a question of subdivision; concealing courage under a show of timidity presupposes a fund of latent energy; masking strength with weakness is to be effected by tactical dispositions.

Thus one who is skilful at keeping the enemy on the move maintains deceitful appearances, according to which the enemy will act. He sacrifices something, that the enemy may snatch at it.

By holding out baits, he keeps him on the march; then with a body of picked men he lies in wait for him.

SEAL SCRIPT

故善戰者，求之於勢，不責於人；故能擇人
而任勢。任勢者，其戰人也，如轉木石。木石之
性，安則靜，危則動，方則止，圓則行。故善戰
人之勢，如轉圓石於千仞之山者，勢也。

故善战者，求之于势，不责于人；故能择人
而任势。任势者，其战人也，如转木石。木石之
性，安则静，危则动，方则止，圆则行。故善战
人之势，如转圆石于千仞之山者，势也。

The clever combatant looks to the effect of combined energy, and does not require too much from individuals. Hence his ability to pick out the right men and utilize combined energy.

When he utilizes combined energy, his fighting men become as it were like unto rolling logs or stones. For it is the nature of a log or stone to remain motionless on level ground, and to move when on a slope; if four-cornered, to come to a standstill, but if round-shaped, to go rolling down.

Thus the energy developed by good fighting men is as the momentum of a round stone rolled down a mountain thousands of feet in height. So much on the subject of energy.

SEAL SCRIPT

虛實

孫子曰：凡先處戰地而待敵者佚，後處戰地而趨戰者勞。故善戰者，致人而不致於人。能使敵人自至者，利之也；能使敵人不得至者，害之也。故敵佚能勞之，飽能饑之，安能動之。出其所必趨，趨其所不意。行千里而不勞者，行於無人之地也；攻而必取者，攻其所不守也。守而必固者，守其所不攻也。

虚实

孙子曰：凡先处战地而待敌者佚，后处战地而趋战者劳。故善战者，致人而不致于人。能使敌人自至者，利之也；能使敌人不得至者，害之也。故敌佚能劳之，饱能饥之，安能动之。出其所必趋，趋其所不意。行千里而不劳者，行于无人之地也；攻而必取者，攻其所不守也。守而必固者，守其所不攻也。

Weak Points and Strong

Sun Tzu said: Whoever is first in the field and awaits the coming of the enemy, will be fresh for the fight; whoever is second in the field and has to hasten to battle will arrive exhausted.

Therefore the clever combatant imposes his will on the enemy, but does not allow the enemy's will to be imposed on him.

By holding out advantages to him, he can cause the enemy to approach of his own accord; or, by inflicting damage, he can make it impossible for the enemy to draw near.

If the enemy is taking his ease, he can harass him; if well supplied with food, he can starve him out; if quietly encamped, he can force him to move.

Appear at points which the enemy must hasten to defend; march swiftly to places where you are not expected.

An army may march great distances without distress, if it marches through country where the enemy is not.

You can be sure of succeeding in your attacks if you only attack places which are undefended. You can ensure the safety of your defence if you only hold positions that cannot be attacked.

SEAL SCRIPT

51

故善攻者，敵不知其所守；善守者，敵不知其所攻。微乎微乎，至於無形；神乎神乎，至於無聲，故能為敵之司命。進而不可禦者，沖其虛也；退而不可追者，速而不可及也。故我欲戰，敵雖高壘深溝，不得不與我戰者，攻其所必救也；我不欲戰，雖畫地而守之，敵不得與我戰者，乖其所之也。

故善攻者，敌不知其所守；善守者，敌不知其所攻。微乎微乎，至于无形；神乎神乎，至于无声，故能为敌之司命。进而不可御者，冲其虚也；退而不可追者，速而不可及也。故我欲战，敌虽高垒深沟，不得不与我战者，攻其所必救也；我不欲战，虽画地而守之，敌不得与我战者，乖其所之也。

Hence that general is skilful in attack whose opponent does not know what to defend; and he is skilful in defence whose opponent does not know what to attack.

O divine art of subtlety and secrecy! Through you we learn to be invisible, through you inaudible; and hence we can hold the enemy's fate in our hands.

You may advance and be absolutely irresistible, if you make for the enemy's weak points; you may retire and be safe from pursuit if your movements are more rapid than those of the enemy.

If we wish to fight, the enemy can be forced to an engagement even though he be sheltered behind a high rampart and a deep ditch. All we need do is attack some other place that he will be obliged to relieve.

If we do not wish to fight, we can prevent the enemy from engaging us even though the lines of our encampment be merely traced out on the ground. All we need do is to throw something odd and unaccountable in his way.

SEAL SCRIPT

故形人而我無形，則我專而敵分。我專為一，敵分為十，是以十攻其一也。則我眾敵寡，能以眾擊寡者，則吾之所與戰者約矣。

故形人而我无形，则我专而敌分。我专为一，敌分为十，是以十攻其一也。则我众敌寡，能以众击寡者，则吾之所与战者约矣。

By discovering the enemy's dispositions and remaining invisible ourselves, we can keep our forces concentrated, while the enemy's must be divided.

We can form a single united body, while the enemy must split up into fractions. Hence there will be a whole pitted against separate parts of a whole, which means that we shall be many to the enemy's few.

And if we are able thus to attack an inferior force with a superior one, our opponents will be in dire straits.

SEAL SCRIPT

吾所與戰之地不可知，不可知則敵所備者多，敵所備者多，則吾所與戰者寡矣。故備前則後寡，備後則前寡，備左則右寡，備右則左寡，無所不備，則無所不寡。寡者，備人者也；眾者，使人備己者也。

吾所与战之地不可知，不可知则敌所备者多，敌所备者多，则吾所与战者寡矣。故备前则后寡，备后则前寡，备左则右寡，备右则左寡，无所不备，则无所不寡。寡者，备人者也；众者，使人备己者也。

The spot where we intend to fight must not be made known; for then the enemy will have to prepare against a possible attack at several different points; and his forces being thus distributed in many directions, the numbers we shall have to face at any given point will be proportionately few.

For should the enemy strengthen his van, he will weaken his rear; should he strengthen his rear, he will weaken his van; should he strengthen his left, he will weaken his right; should he strengthen his right, he will weaken his left. If he sends reinforcements everywhere, he will everywhere be weak.

Numerical weakness comes from having to prepare against possible attacks; numerical strength, from compelling our adversary to make these preparations against us.

SEAL SCRIPT

　　故知戰之地，知戰之日，則可千里而會戰；
不知戰之地，不知戰日，則左不能救右，右不能
救左，前不能救後，後不能救前，而況遠者數十
裏，近者數裏乎！以吾度之，越人之兵雖多，亦
奚益於勝哉！故曰：勝可為也。敵雖眾，可使無
鬥。

　　故知战之地，知战之日，则可千里而会战；
不知战之地，不知战日，则左不能救右，右不能
救左，前不能救后，后不能救前，而况远者数十
里，近者数里乎！以吾度之，越人之兵虽多，亦
奚益于胜哉！故曰：胜可为也。敌虽众，可使无
斗。

Knowing the place and the time of the coming battle, we may concentrate from the greatest distances in order to fight.

But if neither time nor place be known, then the left wing will be impotent to succor the right, the right equally impotent to succor the left, the van unable to relieve the rear, or the rear to support the van. How much more so if the furthest portions of the army are anything under a hundred li apart, and even the nearest are separated by several li!

Though according to my estimate the soldiers of Yüeh exceed our own in number, that shall advantage them nothing in the matter of victory. I say then that victory can be achieved.

Though the enemy be stronger in numbers, we may prevent him from fighting.

SEAL SCRIPT

故策之而知得失之計，候之而知動靜之理，形之而知死生之地，角之而知有餘不足之處。故形兵之極，至於無形。無形則深間不能窺，智者不能謀。因形而措勝於眾，眾不能知。人皆知我所以勝之形，而莫知吾所以制勝之形。故其戰勝不復，而應形於無窮。

故策之而知得失之计，候之而知动静之理，形之而知死生之地，角之而知有余不足之处。故形兵之极，至于无形。无形则深间不能窥，智者不能谋。因形而措胜于众，众不能知。人皆知我所以胜之形，而莫知吾所以制胜之形。故其战胜不复，而应形于无穷。

Scheme so as to discover his plans and the likelihood of their success.

Rouse him, and learn the principle of his activity or inactivity. Force him to reveal himself, so as to find out his vulnerable spots.

Carefully compare the opposing army with your own, so that you may know where strength is superabundant and where it is deficient.

In making tactical dispositions, the highest pitch you can attain is to conceal them; conceal your dispositions, and you will be safe from the prying of the subtlest spies, from the machinations of the wisest brains.

How victory may be produced for them out of the enemy's own tactics—that is what the multitude cannot comprehend.

All men can see the tactics whereby I conquer, but what none can see is the strategy out of which victory is evolved.

Do not repeat the tactics which have gained you one victory, but let your methods be regulated by the infinite variety of circumstances.

SEAL SCRIPT

夫兵形象水，水之行避高而趨下，兵之形避實而擊虛；水因地而制流，兵因敵而制勝。故兵無常勢，水無常形。能因敵變化而取勝者，謂之神。故五行無常勝，四時無常位，日有短長，月有死生。

夫兵形象水，水之行避高而趋下，兵之形避实而击虚；水因地而制流，兵因敌而制胜。故兵无常势，水无常形。能因敌变化而取胜者，谓之神。故五行无常胜，四时无常位，日有短长，月有死生。

Military tactics are like unto water; for water in its natural course runs away from high places and hastens downwards.

So in war, the way is to avoid what is strong and to strike at what is weak.

Water shapes its course according to the nature of the ground over which it flows; the soldier works out his victory in relation to the foe whom he is facing.

Therefore, just as water retains no constant shape, so in warfare there are no constant conditions.

He who can modify his tactics in relation to his opponent and thereby succeed in winning, may be called a heaven-born captain.

The five elements are not always equally predominant; the four seasons make way for each other in turn. There are short days and long; the moon has its periods of waning and waxing.

SEAL SCRIPT

63

軍爭

孫子曰：凡用兵之法，將受命於君，合軍聚眾，交和而舍，莫難於軍爭。軍爭之難者，以迂為直，以患為利。故迂其途，而誘之以利，後人發，先人至，此知迂直之計者也。軍爭為利，軍爭為危。

军争

孙子曰：凡用兵之法，将受命于君，合军聚众，交和而舍，莫难于军争。军争之难者，以迂为直，以患为利。故迂其途，而诱之以利，后人发，先人至，此知迂直之计者也。军争为利，军争为危。

Manoeuvring

Sun Tzu said: In war, the general receives his commands from the sovereign.

Having collected an army and concentrated his forces, he must blend and harmonize the different elements thereof before pitching his camp.

After that, comes tactical manoeuvring, than which there is nothing more difficult. The difficulty of tactical manoeuvring consists in turning the devious into the direct, and misfortune into gain.

Thus, to take a long and circuitous route, after enticing the enemy out of the way, and though starting after him, to contrive to reach the goal before him, shows knowledge of the artifice of deviation.

Manoeuvring with an army is advantageous; with an undisciplined multitude, most dangerous.

SEAL SCRIPT

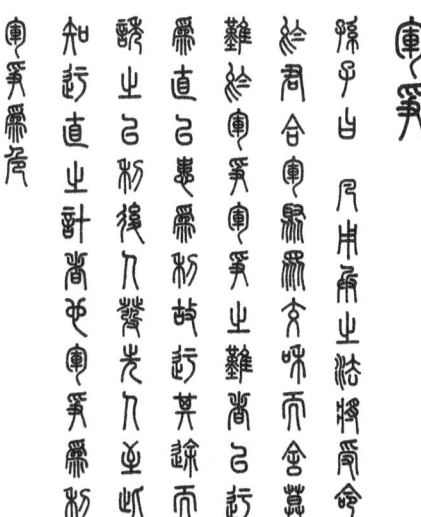

65

舉軍而爭利則不及，委軍而爭利則輜重捐。是故捲甲而趨，日夜不處，倍道兼行，百裡而爭利，則擒三將軍，勁者先，疲者後，其法十一而至；五十裏而爭利，則蹶上將軍，其法半至；三十裏而爭利，則三分之二至。是故軍無輜重則亡，無糧食則亡，無委積則亡。故不知諸侯之謀者，不能豫交；不知山林、險阻、沮澤之形者，不能行軍；不用鄉導者，不能得地利。

举军而争利则不及，委军而争利则辎重捐。是故卷甲而趋，日夜不处，倍道兼行，百里而争利，则擒三将军，劲者先，疲者后，其法十一而至；五十里而争利，则蹶上将军，其法半至；三十里而争利，则三分之二至。是故军无辎重则亡，无粮食则亡，无委积则亡。故不知诸侯之谋者，不能豫交；不知山林、险阻、沮泽之形者，不能行军；不用乡导者，不能得地利。

If you set a fully equipped army in march in order to snatch an advantage, the chances are that you will be too late. On the other hand, to detach a flying column for the purpose involves the sacrifice of its baggage and stores.

Thus, if you order your men to roll up their buff-coats, and make forced marches without halting day or night, covering double the usual distance at a stretch, doing a hundred li in order to wrest an advantage, the leaders of all your three divisions will fall into the hands of the enemy.

The stronger men will be in front, the jaded ones will fall behind, and on this plan only one-tenth of your army will reach its destination.

If you march fifty li in order to outmanoeuvre the enemy, you will lose the leader of your first division, and only half your force will reach the goal.

If you march thirty li with the same object, two-thirds of your army will arrive.

We may take it then that an army without its baggage-train is lost; without provisions it is lost; without bases of supply it is lost.

We cannot enter into alliances until we are acquainted with the designs of our neighbors.

We are not fit to lead an army on the march unless we are familiar with the face of the country—its mountains and forests, its pitfalls and precipices, its marshes and swamps.

SEAL SCRIPT

故兵以詐立，以利動，以分和為變者也。故其疾如風，其徐如林，侵掠如火，不動如山，難知如陰，動如雷震。掠鄉分眾，廓地分利，懸權而動。先知迂直之計者勝，此軍爭之法也。

故兵以诈立，以利动，以分和为变者也。故其疾如风，其徐如林，侵掠如火，不动如山，难知如阴，动如雷震。掠乡分众，廓地分利，悬权而动。先知迂直之计者胜，此军争之法也。

We shall be unable to turn natural advantage to account unless we make use of local guides.

In war, practice dissimulation, and you will succeed. Move only if there is a real advantage to be gained.

Whether to concentrate or to divide your troops, must be decided by circumstances.

Let your rapidity be that of the wind, your compactness that of the forest.

In raiding and plundering be like fire, in immovability like a mountain.

Let your plans be dark and impenetrable as night, and when you move, fall like a thunderbolt.

When you plunder a countryside, let the spoil be divided amongst your men; when you capture new territory, cut it up into allotments for the benefit of the soldiery.

Ponder and deliberate before you make a move.

He will conquer who has learnt the artifice of deviation. Such is the art of manoeuvring.

SEAL SCRIPT

69

軍政曰："言不相聞，故為之金鼓；視不相見，故為之旌旗。"夫金鼓旌旗者，所以一民之耳目也。民既專一，則勇者不得獨進，怯者不得獨退，此用眾之法也。故夜戰多金鼓，晝戰多旌旗，所以變人之耳目也。

军政曰："言不相闻，故为之金鼓；视不相见，故为之旌旗。"夫金鼓旌旗者，所以一民之耳目也。民既专一，则勇者不得独进，怯者不得独退，此用众之法也。故夜战多金鼓，昼战多旌旗，所以变人之耳目也。

The Book of Army Management says: On the field of battle, the spoken word does not carry far enough: hence the institution of gongs and drums. Nor can ordinary objects be seen clearly enough: hence the institution of banners and flags.

Gongs and drums, banners and flags, are means whereby the ears and eyes of the host may be focused on one particular point.

The host thus forming a single united body, is it impossible either for the brave to advance alone, or for the cowardly to retreat alone. This is the art of handling large masses of men.

In night-fighting, then, make much use of signal-fires and drums, and in fighting by day, of flags and banners, as a means of influencing the ears and eyes of your army.

SEAL SCRIPT

多煌旗旛所以變人之耳目也
止怯不得故夜戰多金鼓畫戰
旛難怙肯不得旛揮耵用眾
目色民眾壹所爾肯不得
金鼓煌旗旛所以己一所
鼙視不相見故爲之煌旗木
軍政曰書不相聞故爲之金

三軍可奪氣，將軍可奪心。是故朝氣銳，畫氣惰，暮氣歸。善用兵者，避其銳氣，擊其惰歸，此治氣者也。以治待亂，以靜待嘩，此治心者也。以近待遠，以佚待勞，以飽待饑，此治力者也。

三军可夺气，将军可夺心。是故朝气锐，昼气惰，暮气归。善用兵者，避其锐气，击其惰归，此治气者也。以治待乱，以静待哗，此治心者也。以近待远，以佚待劳，以饱待饥，此治力者也。

A whole army may be robbed of its spirit; a commander-in-chief may be robbed of his presence of mind.

Now a soldier's spirit is keenest in the morning; by noonday it has begun to flag; and in the evening, his mind is bent only on returning to camp.

A clever general, therefore, avoids an army when its spirit is keen, but attacks it when it is sluggish and inclined to return. This is the art of studying moods.

Disciplined and calm, to await the appearance of disorder and hubbub amongst the enemy:—this is the art of retaining self-possession.

To be near the goal while the enemy is still far from it, to wait at ease while the enemy is toiling and struggling, to be well-fed while the enemy is famished:—this is the art of husbanding one's strength.

SEAL SCRIPT

73

　　無邀正正之旗，無擊堂堂之陳，此治變者也。故用兵之法，高陵勿向，背丘勿逆，佯北勿從，銳卒勿攻，餌兵勿食，歸師勿遏，圍師遺闕，窮寇勿迫，此用兵之法也。

　　无邀正正之旗，无击堂堂之陈，此治变者也。故用兵之法，高陵勿向，背丘勿逆，佯北勿从，锐卒勿攻，饵兵勿食，归师勿遏，围师遗阙，穷寇勿迫，此用兵之法也。

To refrain from intercepting an enemy whose banners are in perfect order, to refrain from attacking an army drawn up in calm and confident array:—this is the art of studying circumstances.

It is a military axiom not to advance uphill against the enemy, nor to oppose him when he comes downhill.

Do not pursue an enemy who simulates flight; do not attack soldiers whose temper is keen.

Do not swallow bait offered by the enemy. Do not interfere with an army that is returning home.

When you surround an army, leave an outlet free. Do not press a desperate foe too hard.

Such is the art of warfare.

SEAL SCRIPT

九變

孫子曰：凡用兵之法，將受命於君，合軍聚合。泛地無舍，衢地合交，絕地無留，圍地則謀，死地則戰，途有所不由，軍有所不擊，城有所不攻，地有所不爭，君命有所不受。故將通於九變之利者，知用兵矣；將不通九變之利，雖知地形，不能得地之利矣；治兵不知九變之術，雖知五利，不能得人之用矣。

九变

孙子曰：凡用兵之法，将受命于君，合军聚合。泛地无舍，衢地合交，绝地无留，围地则谋，死地则战，途有所不由，军有所不击，城有所不攻，地有所不争，君命有所不受。故将通于九变之利者，知用兵矣；将不通九变之利，虽知地形，不能得地之利矣；治兵不知九变之术，虽知五利，不能得人之用矣。

Variation of Tactics

Sun Tzu said: In war, the general receives his commands from the sovereign, collects his army and concentrates his forces.

When in difficult country, do not encamp. In country where high roads intersect, join hands with your allies. Do not linger in dangerously isolated positions. In hemmed-in situations, you must resort to stratagem. In a desperate position, you must fight.

There are roads which must not be followed, armies which must not be attacked, towns which must not be besieged, positions which must not be contested, commands of the sovereign which must not be obeyed.

The general who thoroughly understands the advantages that accompany variation of tactics knows how to handle his troops.

The general who does not understand these, may be well acquainted with the configuration of the country, yet he will not be able to turn his knowledge to practical account.

So, the student of war who is unversed in the art of varying his plans, even though he be acquainted with the Five Advantages, will fail to make the best use of his men.

SEAL SCRIPT

77

　　是故智者之慮，必雜於利害，雜於利而務可信也，雜於害而患可解也。是故屈諸侯者以害，役諸侯者以業，趨諸侯者以利。

　　故用兵之法，無恃其不來，恃吾有以待之；無恃其不攻，恃吾有所不可攻也。

　　是故智者之虑，必杂于利害，杂于利而务可信也，杂于害而患可解也。是故屈诸侯者以害，役诸侯者以业，趋诸侯者以利。

　　故用兵之法，无恃其不来，恃吾有以待之；无恃其不攻，恃吾有所不可攻也。

Hence in the wise leader's plans, considerations of advantage and of disadvantage will be blended together.

If our expectation of advantage be tempered in this way, we may succeed in accomplishing the essential part of our schemes.

If, on the other hand, in the midst of difficulties we are always ready to seize an advantage, we may extricate ourselves from misfortune.

Reduce the hostile chiefs by inflicting damage on them; and make trouble for them, and keep them constantly engaged; hold out specious allurements, and make them rush to any given point.

The art of war teaches us to rely not on the likelihood of the enemy's not coming, but on our own readiness to receive him; not on the chance of his not attacking, but rather on the fact that we have made our position unassailable.

SEAL SCRIPT

故將有五危，必死可殺，必生可虜，忿速可
侮，廉潔可辱，愛民可煩。凡此五者，將之過也，
用兵之災也。覆軍殺將，必以五危，不可不察也。

故将有五危，必死可杀，必生可虏，忿速可
侮，廉洁可辱，爱民可烦。凡此五者，将之过也，
用兵之灾也。覆军杀将，必以五危，不可不察也。

There are five dangerous faults which may affect a general:
Recklessness, which leads to destruction;
cowardice, which leads to capture;
a hasty temper, which can be provoked by insults;
a delicacy of honor which is sensitive to shame;
over-solicitude for his men, which exposes him to worry and trouble.
These are the five besetting sins of a general, ruinous to the conduct of war.
When an army is overthrown and its leader slain, the cause will surely be found among these five dangerous faults. Let them be a subject of meditation.

SEAL SCRIPT

行軍

孫子曰：凡處軍相敵，絕山依穀，視生處高，戰隆無登，此處山之軍也。絕水必遠水，客絕水而來，勿迎之於水內，令半渡而擊之利，欲戰者，無附於水而迎客，視生處高，無迎水流，此處水上之軍也。

行军

孙子曰：凡处军相敌，绝山依谷，视生处高，战隆无登，此处山之军也。绝水必远水，客绝水而来，勿迎之于水内，令半渡而击之利，欲战者，无附于水而迎客，视生处高，无迎水流，此处水上之军也。

The Army on the March

Sun Tzu said: We come now to the question of encamping the army, and observing signs of the enemy. Pass quickly over mountains, and keep in the neighborhood of valleys.

Camp in high places, facing the sun. Do not climb heights in order to fight. So much for mountain warfare.

After crossing a river, you should get far away from it.

When an invading force crosses a river in its onward march, do not advance to meet it in midstream. It will be best to let half the army get across, and then deliver your attack.

If you are anxious to fight, you should not go to meet the invader near a river which he has to cross.

Moor your craft higher up than the enemy, and facing the sun. Do not move upstream to meet the enemy. So much for river warfare.

SEAL SCRIPT

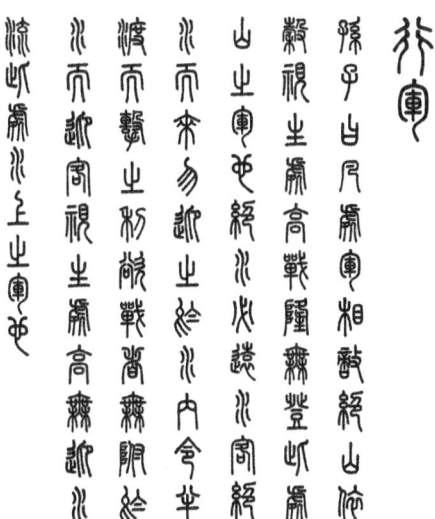

83

絕斥澤，唯亟去無留，若交軍於斥澤之中，必依水草而背眾樹，此處斥澤之軍也。平陸處易，右背高，前死後生，此處平陸之軍也。凡此四軍之利，黃帝之所以勝四帝也。

绝斥泽，唯亟去无留，若交军于斥泽之中，必依水草而背众树，此处斥泽之军也。平陆处易，右背高，前死后生，此处平陆之军也。凡此四军之利，黄帝之所以胜四帝也。

In crossing salt-marshes, your sole concern should be to get over them quickly, without any delay.

If forced to fight in a salt-marsh, you should have water and grass near you, and get your back to a clump of trees. So much for operations in salt-marshes.

In dry, level country, take up an easily accessible position with rising ground to your right and on your rear, so that the danger may be in front, and safety lie behind. So much for campaigning in flat country.

These are the four useful branches of military knowledge which enabled the Yellow Emperor to vanquish four several sovereigns.

SEAL SCRIPT

凡軍好高而惡下，貴陽而賤陰，養生而處實，軍無百疾，是謂必勝。丘陵堤防，必處其陽而右背之，此兵之利，地之助也。上雨水流至，欲涉者，待其定也。凡地有絕澗、天井、天牢、天羅、天陷、天隙，必亟去之，勿近也。吾遠之，敵近之；吾迎之，敵背之。軍旁有險阻、潢井、蒹葭、小林、蘙薈者，必謹覆索之，此伏姦之所處也。

凡军好高而恶下，贵阳而贱阴，养生而处实，军无百疾，是谓必胜。丘陵堤防，必处其阳而右背之，此兵之利，地之助也。上雨水流至，欲涉者，待其定也。凡地有绝涧、天井、天牢、天罗、天陷、天隙，必亟去之，勿近也。吾远之，敌近之；吾迎之，敌背之。军旁有险阻、潢井、蒹葭、小林、蘙荟者，必谨覆索之，此伏奸之所处也。

All armies prefer high ground to low, and sunny places to dark.

If you are careful of your men, and camp on hard ground, the army will be free from disease of every kind, and this will spell victory.

When you come to a hill or a bank, occupy the sunny side, with the slope on your right rear. Thus you will at once act for the benefit of your soldiers and utilize the natural advantages of the ground.

When, in consequence of heavy rains upcountry, a river which you wish to ford is swollen and flecked with foam, you must wait until it subsides.

Country in which there are precipitous cliffs with torrents running between, deep natural hollows, confined places, tangled thickets, quagmires and crevasses, should be left with all possible speed and not approached.

While we keep away from such places, we should get the enemy to approach them; while we face them, we should let the enemy have them on his rear.

If in the neighborhood of your camp there should be any hilly country, ponds surrounded by aquatic grass, hollow basins filled with reeds, or woods with thick undergrowth, they must be carefully routed out and searched; for these are places where men in ambush or insidious spies are likely to be lurking.

SEAL SCRIPT

敵近而靜者，恃其險也；遠而挑戰者，欲人
之進也；其所居易者，利也；眾樹動者，來也；
眾草多障者，疑也；鳥起者，伏也；獸駭者，覆也；
塵高而銳者，車來也；卑而廣者，徒來也；散而
條達者，樵採也；少而往來者，營軍也；

敌近而静者，恃其险也；远而挑战者，欲人
之进也；其所居易者，利也；众树动者，来也；
众草多障者，疑也；鸟起者，伏也；兽骇者，覆也；
尘高而锐者，车来也；卑而广者，徒来也；散而
条达者，樵采也；少而往来者，营军也；

When the enemy is close at hand and remains quiet, he is relying on the natural strength of his position.

When he keeps aloof and tries to provoke a battle, he is anxious for the other side to advance.

If his place of encampment is easy of access, he is tendering a bait.

Movement amongst the trees of a forest shows that the enemy is advancing. The appearance of a number of screens in the midst of thick grass means that the enemy wants to make us suspicious.

The rising of birds in their flight is the sign of an ambuscade. Startled beasts indicate that a sudden attack is coming.

When there is dust rising in a high column, it is the sign of chariots advancing; when the dust is low, but spread over a wide area, it betokens the approach of infantry. When it branches out in different directions, it shows that parties have been sent to collect firewood. A few clouds of dust moving to and fro signify that the army is encamping.

SEAL SCRIPT

辭卑而備者，進也；辭強而進驅者，退也；輕車先出居其側者，陳也；無約而請和者，謀也；奔走而陳兵者，期也；半進半退者，誘也；杖而立者，饑也；汲而先飲者，渴也；見利而不進者，勞也；鳥集者，虛也；夜呼者，恐也；軍擾者，將不重也；旌旗動者，亂也；吏怒者，倦也；

辞卑而备者，进也；辞强而进驱者，退也；轻车先出居其侧者，陈也；无约而请和者，谋也；奔走而陈兵者，期也；半进半退者，诱也；杖而立者，饥也；汲而先饮者，渴也；见利而不进者，劳也；鸟集者，虚也；夜呼者，恐也；军扰者，将不重也；旌旗动者，乱也；吏怒者，倦也；

Humble words and increased preparations are signs that the enemy is about to advance. Violent language and driving forward as if to the attack are signs that he will retreat.

When the light chariots come out first and take up a position on the wings, it is a sign that the enemy is forming for battle.

Peace proposals unaccompanied by a sworn covenant indicate a plot.

When there is much running about and the soldiers fall into rank, it means that the critical moment has come.

When some are seen advancing and some retreating, it is a lure.

When the soldiers stand leaning on their spears, they are faint from want of food.

If those who are sent to draw water begin by drinking themselves, the army is suffering from thirst.

If the enemy sees an advantage to be gained and makes no effort to secure it, the soldiers are exhausted.

If birds gather on any spot, it is unoccupied. Clamor by night betokens nervousness.

If there is disturbance in the camp, the general's authority is weak. If the banners and flags are shifted about, sedition is afoot. If the officers are angry, it means that the men are weary.

SEAL SCRIPT

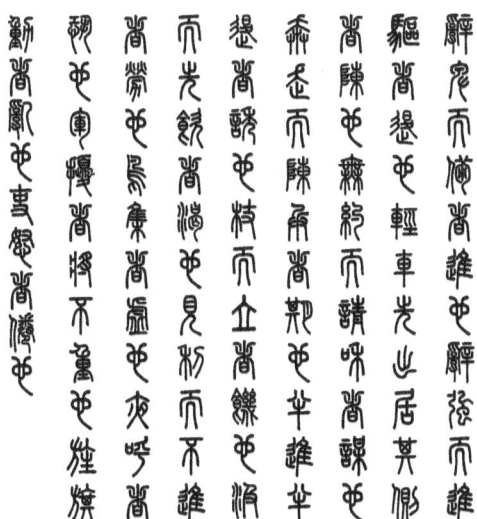

殺馬肉食者，軍無糧也；懸甀不返其舍者，窮寇也；諄諄翕翕，徐與人言者，失眾也；數賞者，窘也；數罰者，困也；先暴而後畏其眾者，不精之至也；來委謝者，欲休息也。兵怒而相迎，久而不合，又不相去，必謹察之。

杀马肉食者，军无粮也；悬甀不返其舍者，穷寇也；谆谆翕翕，徐与人言者，失众也；数赏者，窘也；数罚者，困也；先暴而后畏其众者，不精之至也；来委谢者，欲休息也。兵怒而相迎，久而不合，又不相去，必谨察之。

When an army feeds its horses with grain and kills its cattle for food, and when the men do not hang their cooking-pots over the campfires, showing that they will not return to their tents, you may know that they are determined to fight to the death.

The sight of men whispering together in small knots or speaking in subdued tones points to disaffection amongst the rank and file.

Too frequent rewards signify that the enemy is at the end of his resources; too many punishments betray a condition of dire distress.

To begin by bluster, but afterwards to take fright at the enemy's numbers, shows a supreme lack of intelligence.

When envoys are sent with compliments in their mouths, it is a sign that the enemy wishes for a truce.

If the enemy's troops march up angrily and remain facing ours for a long time without either joining battle or taking themselves off again, the situation is one that demands great vigilance and circumspection.

SEAL SCRIPT

兵非貴益多也，惟無武進，足以並力料敵取人而已。夫惟無慮而易敵者，必擒於人。卒未親而罰之，則不服，不服則難用。卒已親附而罰不行，則不可用。故合之以文，齊之以武，是謂必取。令素行以教其民，則民服；令素不行以教其民，則民不服。令素行者，與眾相得也。

兵非贵益多也，惟无武进，足以并力料敌取人而已。夫惟无虑而易敌者，必擒于人。卒未亲而罚之，则不服，不服则难用。卒已亲附而罚不行，则不可用。故合之以文，齐之以武，是谓必取。令素行以教其民，则民服；令素不行以教其民，则民不服。令素行者，与众相得也。

If our troops are no more in number than the enemy, that is amply sufficient; it only means that no direct attack can be made. What we can do is simply to concentrate all our available strength, keep a close watch on the enemy, and obtain reinforcements.

He who exercises no forethought but makes light of his opponents is sure to be captured by them.

If soldiers are punished before they have grown attached to you, they will not prove submissive; and, unless submissive, then will be practically useless. If, when the soldiers have become attached to you, punishments are not enforced, they will still be useless.

Therefore soldiers must be treated in the first instance with humanity, but kept under control by means of iron discipline. This is a certain road to victory.

If in training soldiers commands are habitually enforced, the army will be well-disciplined; if not, its discipline will be bad.

If a general shows confidence in his men but always insists on his orders being obeyed, the gain will be mutual.

SEAL SCRIPT

地形

孫子曰：地形有通者、有掛者、有支者、有隘者、有險者、有遠者。我可以往，彼可以來，曰通。通形者，先居高陽，利糧道，以戰則利。

地形

孙子曰：地形有通者、有挂者、有支者、有隘者、有险者、有远者。我可以往，彼可以来，曰通。通形者，先居高阳，利粮道，以战则利。

Terrain

Sun Tzu said: We may distinguish six kinds of terrain, to wit: (1) Accessible ground; (2) entangling ground; (3) temporizing ground; (4) narrow passes; (5) precipitous heights; (6) positions at a great distance from the enemy.

Ground which can be freely traversed by both sides is called accessible.

With regard to ground of this nature, be before the enemy in occupying the raised and sunny spots, and carefully guard your line of supplies. Then you will be able to fight with advantage.

SEAL SCRIPT

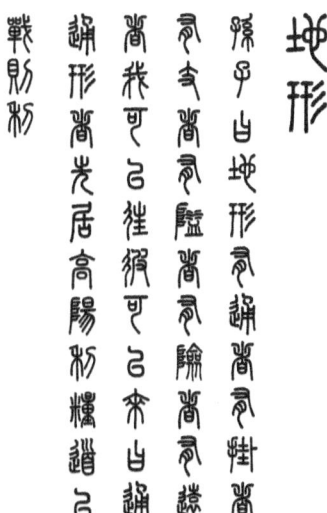

可以往，難以返，曰掛。掛形者，敵無備，出而勝之，敵若有備，出而不勝，難以返，不利。我出而不利，彼出而不利，曰支。支形者，敵雖利我，我無出也，引而去之，令敵半出而擊之利。

可以往，难以返，曰挂。挂形者，敌无备，出而胜之，敌若有备，出而不胜，难以返，不利。我出而不利，彼出而不利，曰支。支形者，敌虽利我，我无出也，引而去之，令敌半出而击之利。

Ground which can be abandoned but is hard to re-occupy is called entangling.

From a position of this sort, if the enemy is unprepared, you may sally forth and defeat him. But if the enemy is prepared for your coming, and you fail to defeat him, then, return being impossible, disaster will ensue.

When the position is such that neither side will gain by making the first move, it is called temporizing ground.

In a position of this sort, even though the enemy should offer us an attractive bait, it will be advisable not to stir forth, but rather to retreat, thus enticing the enemy in his turn; then, when part of his army has come out, we may deliver our attack with advantage.

SEAL SCRIPT

99

隘形者，我先居之，必盈之以待敵。若敵先居之，盈而勿從，不盈而從之。險形者，我先居之，必居高陽以待敵；若敵先居之，引而去之，勿從也。遠形者，勢均難以挑戰，戰而不利。凡此六者，地之道也，將之至任，不可不察也。

隘形者，我先居之，必盈之以待敌。若敌先居之，盈而勿从，不盈而从之。险形者，我先居之，必居高阳以待敌；若敌先居之，引而去之，勿从也。远形者，势均难以挑战，战而不利。凡此六者，地之道也，将之至任，不可不察也。

With regard to narrow passes, if you can occupy them first, let them be strongly garrisoned and await the advent of the enemy.

Should the army forestall you in occupying a pass, do not go after him if the pass is fully garrisoned, but only if it is weakly garrisoned.

With regard to precipitous heights, if you are beforehand with your adversary, you should occupy the raised and sunny spots, and there wait for him to come up.

If the enemy has occupied them before you, do not follow him, but retreat and try to entice him away.

If you are situated at a great distance from the enemy, and the strength of the two armies is equal, it is not easy to provoke a battle, and fighting will be to your disadvantage.

These six are the principles connected with Earth. The general who has attained a responsible post must be careful to study them.

SEAL SCRIPT

凡兵有走者、有馳者、有陷者、有崩者、有亂者、有北者。凡此六者，非天地之災，將之過也。夫勢均，以一擊十，曰走;卒強吏弱，曰馳;吏強卒弱，曰陷;大吏怒而不服，遇敵懟而自戰，將不知其能，曰崩；將弱不嚴，教道不明，吏卒無常，陳兵縱橫，曰亂;將不能料敵，以少合眾，以弱擊強，兵無選鋒，曰北。凡此六者，敗之道也，將之至任，不可不察也。

凡兵有走者、有驰者、有陷者、有崩者、有乱者、有北者。凡此六者，非天地之灾，将之过也。夫势均，以一击十，曰走;卒强吏弱，曰驰;吏强卒弱，曰陷;大吏怒而不服，遇敌怼而自战，将不知其能，曰崩；将弱不严，教道不明，吏卒无常，陈兵纵横，曰乱;将不能料敌，以少合众，以弱击强，兵无选锋，曰北。凡此六者，败之道也，将之至任，不可不察也。

Now an army is exposed to six several calamities, not arising from natural causes, but from faults for which the general is responsible. These are: (1) Flight; (2) insubordination; (3) collapse; (4) ruin; (5) disorganization; (6) rout.

Other conditions being equal, if one force is hurled against another ten times its size, the result will be the flight of the former.

When the common soldiers are too strong and their officers too weak, the result is insubordination. When the officers are too strong and the common soldiers too weak, the result is collapse.

When the higher officers are angry and insubordinate, and on meeting the enemy give battle on their own account from a feeling of resentment, before the commander-in-chief can tell whether or no he is in a position to fight, the result is ruin.

When the general is weak and without authority; when his orders are not clear and distinct; when there are no fixed duties assigned to officers and men, and the ranks are formed in a slovenly haphazard manner, the result is utter disorganization.

When a general, unable to estimate the enemy's strength, allows an inferior force to engage a larger one, or hurls a weak detachment against a powerful one, and neglects to place picked soldiers in the front rank, the result must be a rout.

These are six ways of courting defeat, which must be carefully noted by the

SEAL SCRIPT

103

夫地形者，兵之助也。料敵制勝，計險隘遠近，上將之道也。知此而用戰者必勝，不知此而用戰者必敗。故戰道必勝，主曰無戰，必戰可也；戰道不勝，主曰必戰，無戰可也。

夫地形者，兵之助也。料敌制胜，计险隘远近，上将之道也。知此而用战者必胜，不知此而用战者必败。故战道必胜，主曰无战，必战可也；战道不胜，主曰必战，无战可也。

general who has attained a responsible post.

The natural formation of the country is the soldier's best ally; but a power of estimating the adversary, of controlling the forces of victory, and of shrewdly calculating difficulties, dangers and distances, constitutes the test of a great general.

He who knows these things, and in fighting puts his knowledge into practice, will win his battles. He who knows them not, nor practices them, will surely be defeated.

If fighting is sure to result in victory, then you must fight, even though the ruler forbid it; if fighting will not result in victory, then you must not fight even at the ruler's bidding.

SEAL SCRIPT

故進不求名，退不避罪，唯民是保，而利於主，國之寶也。視卒如嬰兒，故可以與之赴深溪；視卒如愛子，故可與之俱死。厚而不能使，愛而不能令，亂而不能治，譬若驕子，不可用也。

故进不求名，退不避罪，唯民是保，而利于主，国之宝也。视卒如婴儿，故可以与之赴深溪；视卒如爱子，故可与之俱死。厚而不能使，爱而不能令，乱而不能治，譬若骄子，不可用也。

The general who advances without coveting fame and retreats without fearing disgrace, whose only thought is to protect his country and do good service for his sovereign, is the jewel of the kingdom.

Regard your soldiers as your children, and they will follow you into the deepest valleys; look upon them as your own beloved sons, and they will stand by you even unto death.

If, however, you are indulgent, but unable to make your authority felt; kindhearted, but unable to enforce your commands; and incapable, moreover, of quelling disorder: then your soldiers must be likened to spoilt children; they are useless for any practical purpose.

SEAL SCRIPT

知吾卒之可以擊，而不知敵之不可擊，勝之半也；知敵之可擊，而不知吾卒之不可以擊，勝之半也；知敵之可擊，知吾卒之可以擊，而不知地形之不可以戰，勝之半也。故知兵者，動而不迷，舉而不窮。故曰：知彼知己，勝乃不殆；知天知地，勝乃可全。

知吾卒之可以击，而不知敌之不可击，胜之半也；知敌之可击，而不知吾卒之不可以击，胜之半也；知敌之可击，知吾卒之可以击，而不知地形之不可以战，胜之半也。故知兵者，动而不迷，举而不穷。故曰：知彼知己，胜乃不殆；知天知地，胜乃可全。

If we know that our own men are in a condition to attack, but are unaware that the enemy is not open to attack, we have gone only halfway towards victory.

If we know that the enemy is open to attack, but are unaware that our own men are not in a condition to attack, we have gone only halfway towards victory.

If we know that the enemy is open to attack, and also know that our men are in a condition to attack, but are unaware that the nature of the ground makes fighting impracticable, we have still gone only halfway towards victory.

Hence the experienced soldier, once in motion, is never bewildered; once he has broken camp, he is never at a loss.

Hence the saying: If you know the enemy and know yourself, your victory will not stand in doubt; if you know Heaven and know Earth, you may make your victory complete.

SEAL SCRIPT

九地

孫子曰：用兵之法，有散地，有輕地，有爭地，有交地，有衝地，有重地，有泛地，有圍地，有死地。諸侯自戰其地者，為散地；入人之地不深者，為輕地；我得亦利，彼得亦利者，為爭地；我可以往，彼可以來者，為交地；諸侯之地三屬，先至而得天下眾者，為衝地；入人之地深，背城邑多者，為重地；山林、險阻、沮澤，凡難行之道

九地

孙子曰：用兵之法，有散地，有轻地，有争地，有交地，有衢地，有重地，有泛地，有围地，有死地。诸侯自战其地者，为散地；入人之地不深者，为轻地；我得亦利，彼得亦利者，为争地；我可以往，彼可以来者，为交地；诸侯之地三属，先至而得天下众者，为衢地；入人之地深，背城邑多者，为重地；山林、险阻、沮泽，凡难行之道者，为泛地；所由入者隘，所从归者迂，彼寡可

The Nine Situations

Sun Tzu said: The art of war recognizes nine varieties of ground: (1) Dispersive ground; (2) facile ground; (3) contentious ground; (4) open ground; (5) ground of intersecting highways; (6) serious ground; (7) difficult ground; (8) hemmed-in ground; (9) desperate ground.

When a chieftain is fighting in his own territory, it is dispersive ground.

When he has penetrated into hostile territory, but to no great distance, it is facile ground.

Ground the possession of which imports great advantage to either side, is contentious ground.

Ground on which each side has liberty of movement is open ground.

Ground which forms the key to three contiguous states, so that he who occupies it first has most of the Empire at his command, is a ground of intersecting highways.

When an army has penetrated into the heart of a hostile country, leaving a number of fortified cities in its rear, it is serious ground.

Mountain forests, rugged steeps, marshes and fens—all country that is hard to traverse: this is difficult ground.

Ground which is reached through narrow gorges, and from which we can

SEAL SCRIPT

者，為泛地；所由入者隘，所從歸者迂，彼寡可
以擊吾之眾者，為圍地；疾戰則存，不疾戰則亡
者，為死地。是故散地則無戰，輕地則無止，爭
地則無攻，交地則無絕，衢地則合交，重地則掠，
泛地則行，圍地則謀，死地則戰。

以击吾之众者，为围地；疾战则存，不疾战则亡
者，为死地。是故散地则无战，轻地则无止，争
地则无攻，交地则无绝，衢地则合交，重地则掠，
泛地则行，围地则谋，死地则战。

only retire by tortuous paths, so that a small number of the enemy would suffice to crush a large body of our men: this is hemmed in ground.

Ground on which we can only be saved from destruction by fighting without delay, is desperate ground.

On dispersive ground, therefore, fight not. On facile ground, halt not. On contentious ground, attack not.

On open ground, do not try to block the enemy's way. On the ground of intersecting highways, join hands with your allies.

On serious ground, gather in plunder. In difficult ground, keep steadily on the march.

On hemmed-in ground, resort to stratagem. On desperate ground, fight.

SEAL SCRIPT

古之善用兵者，能使敵人前後不相及，眾寡不相恃，貴賤不相救，上下不相收，卒離而不集，兵合而不齊。合於利而動，不合於利而止。敢問敵眾而整將來，待之若何曰：先奪其所愛則聽矣。兵之情主速，乘人之不及。由不虞之道，攻其所不戒也。

古之善用兵者，能使敌人前后不相及，众寡不相恃，贵贱不相救，上下不相收，卒离而不集，兵合而不齐。合于利而动，不合于利而止。敢问敌众而整将来，待之若何曰：先夺其所爱则听矣。兵之情主速，乘人之不及。由不虞之道，攻其所不戒也。

Those who were called skilful leaders of old knew how to drive a wedge between the enemy's front and rear; to prevent cooperation between his large and small divisions; to hinder the good troops from rescuing the bad, the officers from rallying their men.

When the enemy's men were scattered, they prevented them from concentrating; even when their forces were united, they managed to keep them in disorder.

When it was to their advantage, they made a forward move; when otherwise, they stopped still.

If asked how to cope with a great host of the enemy in orderly array and on the point of marching to the attack, I should say: "Begin by seizing something which your opponent holds dear; then he will be amenable to your will."

Rapidity is the essence of war: take advantage of the enemy's unreadiness, make your way by unexpected routes, and attack unguarded spots.

SEAL SCRIPT

凡為客之道，深入則專。主人不克，掠於饒野，三軍足食。謹養而勿勞，並氣積力，運兵計謀，為不可測。投之無所往，死且不北。死焉不得，士人盡力。兵士甚陷則不懼，無所往則固，深入則拘，不得已則鬥。是故其兵不修而戒，不求而得，不約而親，不令而信，禁祥去疑，至死無所之。

凡为客之道，深入则专。主人不克，掠于饶野，三军足食。谨养而勿劳，并气积力，运兵计谋，为不可测。投之无所往，死且不北。死焉不得，士人尽力。兵士甚陷则不惧，无所往则固，深入则拘，不得已则斗。是故其兵不修而戒，不求而得，不约而亲，不令而信，禁祥去疑，至死无所之。

The following are the principles to be observed by an invading force: The further you penetrate into a country, the greater will be the solidarity of your troops, and thus the defenders will not prevail against you.

Make forays in fertile country in order to supply your army with food.

Carefully study the well-being of your men, and do not overtax them. Concentrate your energy and hoard your strength. Keep your army continually on the move, and devise unfathomable plans.

Throw your soldiers into positions whence there is no escape, and they will prefer death to flight. If they will face death, there is nothing they may not achieve. Officers and men alike will put forth their uttermost strength.

Soldiers when in desperate straits lose the sense of fear. If there is no place of refuge, they will stand firm. If they are in hostile country, they will show a stubborn front. If there is no help for it, they will fight hard.

Thus, without waiting to be marshaled, the soldiers will be constantly on the qui vive; without waiting to be asked, they will do your will; without restrictions, they will be faithful; without giving orders, they can be trusted.

Prohibit the taking of omens, and do away with superstitious doubts. Then, until death itself comes, no calamity need be feared.

SEAL SCRIPT

　　吾士無餘財，非惡貨也；無餘命，非惡壽也。令發之日，士卒坐者涕沾襟，偃臥者涕交頤，投之無所往，諸、劌之勇也。

　　吾士无余财，非恶货也；无余命，非恶寿也。令发之日，士卒坐者涕沾襟，偃卧者涕交颐，投之无所往，诸、刿之勇也。

If our soldiers are not overburdened with money, it is not because they have a distaste for riches; if their lives are not unduly long, it is not because they are disinclined to longevity.

On the day they are ordered out to battle, your soldiers may weep, those sitting up bedewing their garments, and those lying down letting the tears run down their cheeks. But let them once be brought to bay, and they will display the courage of a Chu or a Kuei.

SEAL SCRIPT

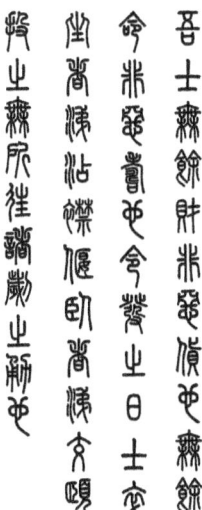

　　故善用兵者，譬如率然。率然者，常山之蛇也。擊其首則尾至，擊其尾則首至，擊其中則首尾俱至。敢問兵可使如率然乎？曰可。夫吳人與越人相惡也，當其同舟而濟而遇風，其相救也如左右手。是故方馬埋輪，未足恃也；齊勇如一，政之道也；剛柔皆得，地之理也。故善用兵者，攜手若使一人，不得已也。

　　故善用兵者，譬如率然。率然者，常山之蛇也。击其首则尾至，击其尾则首至，击其中则首尾俱至。敢问兵可使如率然乎？曰可。夫吴人与越人相恶也，当其同舟而济而遇风，其相救也如左右手。是故方马埋轮，未足恃也；齐勇如一，政之道也；刚柔皆得，地之理也。故善用兵者，携手若使一人，不得已也。

The skilful tactician may be likened to the shuai-jan. Now the shuai-jan is a snake that is found in the Ch'ang mountains. Strike at its head, and you will be attacked by its tail; strike at its tail, and you will be attacked by its head; strike at its middle, and you will be attacked by head and tail both.

Asked if an army can be made to imitate the shuai-jan, I should answer, Yes. For the men of Wu and the men of Yüeh are enemies; yet if they are crossing a river in the same boat and are caught by a storm, they will come to each other's assistance just as the left hand helps the right.

Hence it is not enough to put one's trust in the tethering of horses, and the burying of chariot wheels in the ground.

The principle on which to manage an army is to set up one standard of courage which all must reach.

How to make the best of both strong and weak—that is a question involving the proper use of ground.

Thus the skilful general conducts his army just as though he were leading a single man, willy-nilly, by the hand.

SEAL SCRIPT

121

將軍之事，靜以幽，正以治，能愚士卒之耳目，使之無知；易其事，革其謀，使人無識；易其居，迂其途，使民不得慮。帥與之期，如登高而去其梯；帥與之深入諸侯之地，而發其機。若驅群羊，驅而往，驅而來，莫知所之。聚三軍之眾，投之於險，此謂將軍之事也。九地之變，屈伸之力，人情之理，不可不察也。

将军之事，静以幽，正以治，能愚士卒之耳目，使之无知；易其事，革其谋，使人无识；易其居，迂其途，使民不得虑。帅与之期，如登高而去其梯；帅与之深入诸侯之地，而发其机。若驱群羊，驱而往，驱而来，莫知所之。聚三军之众，投之于险，此谓将军之事也。九地之变，屈伸之力，人情之理，不可不察也。

It is the business of a general to be quiet and thus ensure secrecy; upright and just, and thus maintain order.

He must be able to mystify his officers and men by false reports and appearances, and thus keep them in total ignorance.

By altering his arrangements and changing his plans, he keeps the enemy without definite knowledge. By shifting his camp and taking circuitous routes, he prevents the enemy from anticipating his purpose.

At the critical moment, the leader of an army acts like one who has climbed up a height and then kicks away the ladder behind him. He carries his men deep into hostile territory before he shows his hand.

He burns his boats and breaks his cooking-pots; like a shepherd driving a flock of sheep, he drives his men this way and that, and none knows whither he is going.

To muster his host and bring it into danger:—this may be termed the business of the general.

The different measures suited to the nine varieties of ground; the expediency of aggressive or defensive tactics; and the fundamental laws of human nature: these are things that must most certainly be studied.

SEAL SCRIPT

將軍之事，靜以幽，正以治。能愚士卒之耳目，使之無知。易其事，革其謀，使人無識；易其居，迂其途，使民不得慮。帥與之期，如登高而去其梯；帥與之深入諸侯之地，而發其機。焚舟破釜，若驅群羊，驅而往，驅而來，莫知所之。聚三軍之眾，投之於險，此謂將軍之事也。九地之變，屈伸之利，人情之理，不可不察也。

凡為客之道，深則專，淺則散。去國越境而師者，絕地也；四徹者，衢地也；入深者，重地也；入淺者，輕地也；背固前隘者，圍地也；無所往者，死地也。

凡为客之道，深则专，浅则散。去国越境而师者，绝地也；四彻者，衢地也；入深者，重地也；入浅者，轻地也；背固前隘者，围地也；无所往者，死地也。

When invading hostile territory, the general principle is, that penetrating deeply brings cohesion; penetrating but a short way means dispersion.

When you leave your own country behind, and take your army across neighborhood territory, you find yourself on critical ground. When there are means of communication on all four sides, the ground is one of intersecting highways.

When you penetrate deeply into a country, it is serious ground. When you penetrate but a little way, it is facile ground.

When you have the enemy's strongholds on your rear, and narrow passes in front, it is hemmed-in ground. When there is no place of refuge at all, it is desperate ground.

SEAL SCRIPT

125

是故散地吾將一其志，輕地吾將使之屬，爭
地吾將趨其後，交地吾將謹其守，交地吾將固其
結，衢地吾將謹其恃，重地吾將繼其食，泛地吾
將進其途，圍地吾將塞其闕，死地吾將示之以不
活。故兵之情：圍則禦，不得已則鬥，過則從。

是故散地吾将一其志，轻地吾将使之属，争
地吾将趋其后，交地吾将谨其守，交地吾将固其
结，衢地吾将谨其恃，重地吾将继其食，泛地吾
将进其途，围地吾将塞其阙，死地吾将示之以不
活。故兵之情：围则御，不得已则斗，过则从。

Therefore, on dispersive ground, I would inspire my men with unity of purpose. On facile ground, I would see that there is close connection between all parts of my army.

On contentious ground, I would hurry up my rear.

On open ground, I would keep a vigilant eye on my defences. On ground of intersecting highways, I would consolidate my alliances.

On serious ground, I would try to ensure a continuous stream of supplies. On difficult ground, I would keep pushing on along the road.

On hemmed-in ground, I would block any way of retreat. On desperate ground, I would proclaim to my soldiers the hopelessness of saving their lives.

For it is the soldier's disposition to offer an obstinate resistance when surrounded, to fight hard when he cannot help himself, and to obey promptly when he has fallen into danger.

SEAL SCRIPT

是故不知諸侯之謀者，不能預交；不知山林、險阻、沮澤之形者，不能行軍；不用鄉導，不能得地利。四五者，一不知，非霸王之兵也。

是故不知诸侯之谋者，不能预交；不知山林、险阻、沮泽之形者，不能行军；不用乡导，不能得地利。四五者，一不知，非霸王之兵也。

We cannot enter into alliance with neighboring princes until we are acquainted with their designs. We are not fit to lead an army on the march unless we are familiar with the face of the country—its mountains and forests, its pitfalls and precipices, its marshes and swamps. We shall be unable to turn natural advantages to account unless we make use of local guides.

To be ignorant of any one of the following four or five principles does not befit a warlike prince.

SEAL SCRIPT

夫霸王之兵，伐大國，則其眾不得聚；威加於敵，則其交不得合。是故不爭天下之交，不養天下之權，信己之私，威加於敵，則其城可拔，其國可隳。施無法之賞，懸無政之令。犯三軍之眾，若使一人。犯之以事，勿告以言；犯之以害，勿告以利。投之亡地然後存，陷之死地然後生。夫眾陷於害，然後能為勝敗。

夫霸王之兵，伐大国，则其众不得聚；威加于敌，则其交不得合。是故不争天下之交，不养天下之权，信己之私，威加于敌，则其城可拔，其国可隳。施无法之赏，悬无政之令。犯三军之众，若使一人。犯之以事，勿告以言；犯之以害，勿告以利。投之亡地然后存，陷之死地然后生。夫众陷于害，然后能为胜败。

When a warlike prince attacks a powerful state, his generalship shows itself in preventing the concentration of the enemy's forces. He overawes his opponents, and their allies are prevented from joining against him.

Hence he does not strive to ally himself with all and sundry, nor does he foster the power of other states. He carries out his own secret designs, keeping his antagonists in awe. Thus he is able to capture their cities and overthrow their kingdoms.

Bestow rewards without regard to rule, issue orders without regard to previous arrangements; and you will be able to handle a whole army as though you had to do with but a single man.

Confront your soldiers with the deed itself; never let them know your design. When the outlook is bright, bring it before their eyes; but tell them nothing when the situation is gloomy.

Place your army in deadly peril, and it will survive; plunge it into desperate straits, and it will come off in safety.

For it is precisely when a force has fallen into harm's way that is capable of striking a blow for victory.

SEAL SCRIPT

故為兵之事，在順詳敵之意，並敵一向，千里殺將，是謂巧能成事。是故政舉之日，夷關折符，無通其使，厲於廊廟之上，以誅其事。敵人開闔，必亟入之，先其所愛，微與之期，踐墨隨敵，以決戰事。是故始如處女，敵人開戶；後如脫兔，敵不及拒。

故为兵之事，在顺详敌之意，并敌一向，千里杀将，是谓巧能成事。是故政举之日，夷关折符，无通其使，厉于廊庙之上，以诛其事。敌人开阖，必亟入之，先其所爱，微与之期，践墨随敌，以决战事。是故始如处女，敌人开户；后如脱兔，敌不及拒。

Success in warfare is gained by carefully accommodating ourselves to the enemy's purpose.

By persistently hanging on the enemy's flank, we shall succeed in the long run in killing the commander-in-chief.

This is called ability to accomplish a thing by sheer cunning.

On the day that you take up your command, block the frontier passes, destroy the official tallies, and stop the passage of all emissaries.

Be stern in the council-chamber, so that you may control the situation.

If the enemy leaves a door open, you must rush in.

Forestall your opponent by seizing what he holds dear, and subtly contrive to time his arrival on the ground.

Walk in the path defined by rule, and accommodate yourself to the enemy until you can fight a decisive battle.

At first, then, exhibit the coyness of a maiden, until the enemy gives you an opening; afterwards emulate the rapidity of a running hare, and it will be too late for the enemy to oppose you.

SEAL SCRIPT

火攻

孫子曰：凡火攻有五：一曰火人，二曰火積，三曰火輜，四曰火庫，五曰火隊。行火必有因，因必素具。發火有時，起火有日。時者，天之燥也。日者，月在箕、壁、翼、軫也。凡此四宿者，風起之日也。

火攻

孙子曰：凡火攻有五：一曰火人，二曰火积，三曰火辎，四曰火库，五曰火队。行火必有因，因必素具。发火有时，起火有日。时者，天之燥也。日者，月在箕、壁、翼、轸也。凡此四宿者，风起之日也。

The Attack by Fire

Sun Tzu said: There are five ways of attacking with fire. The first is to burn soldiers in their camp; the second is to burn stores; the third is to burn baggage trains; the fourth is to burn arsenals and magazines; the fifth is to hurl dropping fire amongst the enemy.

In order to carry out an attack, we must have means available; the material for raising fire should always be kept in readiness.

There is a proper season for making attacks with fire, and special days for starting a conflagration.

The proper season is when the weather is very dry; the special days are those when the moon is in the constellations of the Sieve, the Wall, the Wing or the Crossbar; for these four are all days of rising wind.

SEAL SCRIPT

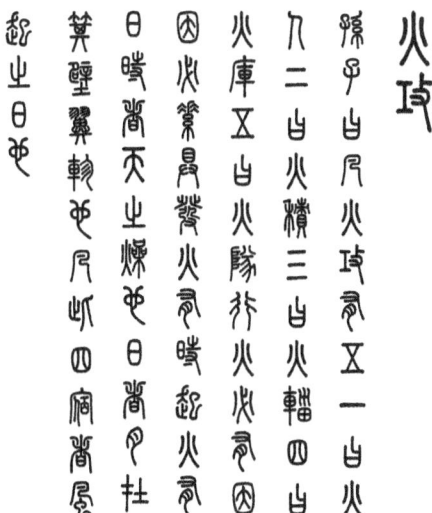

凡火攻，必因五火之變而應之：火發於內，則早應之於外；火發而其兵靜者，待而勿攻，極其火力，可從而從之，不可從則上。火可發於外，無待於內，以時發之，火發上風，無攻下風，畫風久，夜風止。凡軍必知五火之變，以數守之。故以火佐攻者明，以水佐攻者強。水可以絕，不可以奪。

凡火攻，必因五火之变而应之：火发于内，则早应之于外；火发而其兵静者，待而勿攻，极其火力，可从而从之，不可从则上。火可发于外，无待于内，以时发之，火发上风，无攻下风，昼风久，夜风止。凡军必知五火之变，以数守之。故以火佐攻者明，以水佐攻者强。水可以绝，不可以夺。

In attacking with fire, one should be prepared to meet five possible developments:

When fire breaks out inside the enemy's camp, respond at once with an attack from without.

If there is an outbreak of fire, but the enemy's soldiers remain quiet, bide your time and do not attack.

When the force of the flames has reached its height, follow it up with an attack, if that is practicable; if not, stay where you are.

If it is possible to make an assault with fire from without, do not wait for it to break out within, but deliver your attack at a favourable moment.

When you start a fire, be to windward of it. Do not attack from the leeward.

A wind that rises in the daytime lasts long, but a night breeze soon falls.

In every army, the five developments connected with fire must be known, the movements of the stars calculated, and a watch kept for the proper days.

Hence those who use fire as an aid to the attack show intelligence; those who use water as an aid to the attack gain an accession of strength.

By means of water, an enemy may be intercepted, but not robbed of all his belongings.

SEAL SCRIPT

137

夫戰勝攻取而不惰其功者凶，命曰費留。故曰：明主慮之，良將惰之，非利不動，非得不用，非危不戰。主不可以怒而興師，將不可以慍而攻戰。合於利而動，不合於利而上。怒可以複喜，慍可以複說，亡國不可以複存，死者不可以複生。故明主慎之，良將警之。此安國全軍之道也。

夫战胜攻取而不惰其功者凶，命曰"费留"。故曰：明主虑之，良将惰之，非利不动，非得不用，非危不战。主不可以怒而兴师，将不可以慍而攻战。合于利而动，不合于利而上。怒可以复喜，慍可以复说，亡国不可以复存，死者不可以复生。故明主慎之，良将警之。此安国全军之道也。

Unhappy is the fate of one who tries to win his battles and succeed in his attacks without cultivating the spirit of enterprise; for the result is waste of time and general stagnation.

Hence the saying: The enlightened ruler lays his plans well ahead; the good general cultivates his resources.

Move not unless you see an advantage; use not your troops unless there is something to be gained; fight not unless the position is critical.

No ruler should put troops into the field merely to gratify his own spleen; no general should fight a battle simply out of pique.

If it is to your advantage, make a forward move; if not, stay where you are.

Anger may in time change to gladness; vexation may be succeeded by content.

But a kingdom that has once been destroyed can never come again into being; nor can the dead ever be brought back to life.

Hence the enlightened ruler is heedful, and the good general full of caution. This is the way to keep a country at peace and an army intact.

SEAL SCRIPT

139

用間

孫子曰：凡興師十萬，出征千里，百姓之費，公家之奉，日費千金，內外騷動，怠於道路，不得操事者，七十萬家。相守數年，以爭一日之勝，而愛爵祿百金，不知敵之情者，不仁之至也，非民之將也，非主之佐也，非勝之主也。故明君賢將所以動而勝人，成功出於眾者，先知也。先知者，不可取於鬼神，不可象於事，不可驗於度，必取

用间

孙子曰：凡兴师十万，出征千里，百姓之费，公家之奉，日费千金，内外骚动，怠于道路，不得操事者，七十万家。相守数年，以争一日之胜，而爱爵禄百金，不知敌之情者，不仁之至也，非民之将也，非主之佐也，非胜之主也。故明君贤将所以动而胜人，成功出于众者，先知也。先知者，不可取于鬼神，不可象于事，不可验于度，必取于人，知敌之情者也。

The Use of Spies

Sun Tzu said: Raising a host of a hundred thousand men and marching them great distances entails heavy loss on the people and a drain on the resources of the State. The daily expenditure will amount to a thousand ounces of silver. There will be commotion at home and abroad, and men will drop down exhausted on the highways. As many as seven hundred thousand families will be impeded in their labour.

Hostile armies may face each other for years, striving for the victory which is decided in a single day. This being so, to remain in ignorance of the enemy's condition simply because one grudges the outlay of a hundred ounces of silver in honors and emoluments, is the height of inhumanity.

One who acts thus is no leader of men, no present help to his sovereign, no master of victory.

Thus, what enables the wise sovereign and the good general to strike and conquer, and achieve things beyond the reach of ordinary men, is foreknowledge.

Now this foreknowledge cannot be elicited from spirits; it cannot be obtained inductively from experience, nor by any deductive calculation.

Knowledge of the enemy's dispositions can only be obtained from other men.

SEAL SCRIPT

於人，知敵之情者也。

故用間有五：有因間，有內間，有反間，有死間，有生間。五間俱起，莫知其道，是謂神紀，人君之寶也。鄉間者，因其鄉人而用之；內間者，因其官人而用之；反間者，因其敵間而用之；死間者，為誑事於外，令吾聞知之而傳於敵間也；生間者，反報也。

故用间有五：有因间，有内间，有反间，有死间，有生间。五间俱起，莫知其道，是谓神纪，人君之宝也。乡间者，因其乡人而用之；内间者，因其官人而用之；反间者，因其敌间而用之；死间者，为诳事于外，令吾闻知之而传于敌间也；生间者，反报也。

Hence the use of spies, of whom there are five classes: (1) Local spies; (2) inward spies; (3) converted spies; (4) doomed spies; (5) surviving spies.

When these five kinds of spy are all at work, none can discover the secret system. This is called "divine manipulation of the threads." It is the sovereign's most precious faculty.

Having local spies means employing the services of the inhabitants of a district.

Having inward spies, making use of officials of the enemy.

Having converted spies, getting hold of the enemy's spies and using them for our own purposes.

Having doomed spies, doing certain things openly for purposes of deception, and allowing our spies to know of them and report them to the enemy.

Surviving spies, finally, are those who bring back news from the enemy's camp.

SEAL SCRIPT

TRADITIONAL CHINESE

故三軍之事，莫親於間，賞莫厚於間，事莫
密於間，非聖賢不能用間，非仁義不能使間，非
微妙不能得間之實。微哉微哉！無所不用間也。
間事未發而先聞者，間與所告者兼死。

SIMPLIFIED CHINESE

故三军之事，莫亲于间，赏莫厚于间，事莫
密于间，非圣贤不能用间，非仁义不能使间，非
微妙不能得间之实。微哉微哉！无所不用间也。
间事未发而先闻者，间与所告者兼死。

144

Hence it is that which none in the whole army are more intimate relations to be maintained than with spies. None should be more liberally rewarded. In no other business should greater secrecy be preserved.

Spies cannot be usefully employed without a certain intuitive sagacity.

They cannot be properly managed without benevolence and straightforwardness.

Without subtle ingenuity of mind, one cannot make certain of the truth of their reports.

Be subtle! be subtle! and use your spies for every kind of business.

If a secret piece of news is divulged by a spy before the time is ripe, he must be put to death together with the man to whom the secret was told.

SEAL SCRIPT

145

　　凡軍之所欲擊，城之所欲攻，人之所欲殺，
必先知其守將、左右、謁者、門者、舍人之姓名，
令吾間必索知之。敵間之來間我者，因而利之，
導而舍之，故反間可得而用也；因是而知之，故
鄉間、內間可得而使也；因是而知之，故死間為
誑事，可使告敵；因是而知之，故生間可使如期。
五間之事，主必知之，知之必在於反間，故反間
不可不厚也。

　　凡军之所欲击，城之所欲攻，人之所欲杀，
必先知其守将、左右、谒者、门者、舍人之姓名，
令吾间必索知之。敌间之来间我者，因而利之，
导而舍之，故反间可得而用也；因是而知之，故
乡间、内间可得而使也；因是而知之，故死间为
诳事，可使告敌；因是而知之，故生间可使如期。
五间之事，主必知之，知之必在于反间，故反间
不可不厚也。

Whether the object be to crush an army, to storm a city, or to assassinate an individual, it is always necessary to begin by finding out the names of the attendants, the aides-de-camp, and doorkeepers and sentries of the general in command. Our spies must be commissioned to ascertain these.

The enemy's spies who have come to spy on us must be sought out, tempted with bribes, led away and comfortably housed. Thus they will become converted spies and available for our service.

It is through the information brought by the converted spy that we are able to acquire and employ local and inward spies.

It is owing to his information, again, that we can cause the doomed spy to carry false tidings to the enemy.

Lastly, it is by his information that the surviving spy can be used on appointed occasions.

The end and aim of spying in all its five varieties is knowledge of the enemy; and this knowledge can only be derived, in the first instance, from the converted spy. Hence it is essential that the converted spy be treated with the utmost liberality.

SEAL SCRIPT

　　昔殷之興也,伊摯在夏;周之興也,呂牙在殷。故明君賢將，能以上智為間者，必成大功。此兵之要，三軍之所恃而動也。

　　昔殷之兴也,伊挚在夏;周之兴也,吕牙在殷。故明君贤将，能以上智为间者，必成大功。此兵之要，三军之所恃而动也。

Of old, the rise of the Yin dynasty was due to I Chih who had served under the Hsia. Likewise, the rise of the Chou dynasty was due to Lu Ya who had served under the Yin.

Hence it is only the enlightened ruler and the wise general who will use the highest intelligence of the army for purposes of spying and thereby they achieve great results. Spies are a most important element in war, because on them depends an army's ability to move.

SEAL SCRIPT

DISCOVER MORE ANCIENT WISDOM

We invite you to discover another treasure of ancient Chinese wisdom:

"Tao Te Ching" by Lao Tzu

Ultimate Bilingual Edition (4-in-1)

While Sun Tzu taught us the art of strategy, Lao Tzu explores the fundamental nature of existence and living in accord with the Way. This special edition includes the complete text in:

- English
- Traditional Chinese
- Simplified Chinese
- Ancient Seal Script

This unique format lets you explore different translations and witness the evolution of Chinese writing. Whether you're interested in philosophy, language, or personal growth, Lao Tzu's timeless wisdom will speak to you just as powerfully as Sun Tzu's.

Join thousands of readers who have discovered the profound insights of the Tao Te Ching through this comprehensive edition.

Find your copy at major online bookstores worldwide.

www.ingramcontent.com/pod-product-compliance
Lightning Source LLC
Chambersburg PA
CBHW020358130626
46549CB00006B/2339